COPING WITH . . .
WOOD
TRASH

COPING WITH . . .
WOOD TRASH

▾▾▾▾▾▾▾▾▾▾▾▾▾▾▾▾▾▾▾▾

Jamie Daniel • Veronica Bonar
Illustrated by Tony Kenyon

Gareth Stevens Publishing
MILWAUKEE

**For a free color catalog describing Gareth Stevens' list
of high-quality books, call 1-800-341-3569 (USA) or
1-800-461-9120 (Canada).**

Library of Congress Cataloging-in-Publication Data

Daniel, Jamie.
    Coping with— wood trash/adapted from Veronica Bonar's Wood rubbish! by Jamie
Daniel; illustrated by Tony Kenyon. — North American ed.
      p. cm. — (Trash busters)
    Includes bibliographical references and index.
    ISBN 0-8368-1061-9
    1. Wood waste—Juvenile literature. 2. Refuse and refuse disposal—Juvenile literature.
3. Deforestation—Environmental aspects—Juvenile literature. [1. Wood—Recycling. 2.
Refuse and refuse disposal. 3. Recycling (Waste)] I. Kenyon, Tony, ill. II. Bonar,
Veronica. Wood rubbish! III. Title. IV. Series: Daniel, Jamie. Trash busters.
TD810.D36   1994
363.72'82--dc20
                                         93-37686

This North American edition first published in 1994 by

Gareth Stevens Publishing

1555 North RiverCenter Drive, Suite 201
Milwaukee, WI 53212, USA

This edition © 1994 by Zoë Books Limited. First produced as WOOD
RUBBISH!, © 1992 by Zoë Books Limited, original text © 1992 by
Veronica Bonar. Additional end matter © 1994 by Gareth Stevens, Inc.
First published in the USA by arrangement with Zoë Books Limited,
Winchester, England. Published in Canada by arrangement with
Heinemann Educational Books Ltd., Oxford, England.

Series editor: Patricia Lantier-Sampon
Cover design: Karen Knutson

Picture Credits:
The Environmental Picture Library p. 19 (H. Girardet); Robert Harding Picture
Library p. 7; Holt Studios p. 26 (Nigel Cattlin); Oxford Scientific Films pp. 12, 25
(Terry Heathcote), p. 14 (Philippe Henry); Science Photo Library p. 16 (Astrid
and Hanns Frieder Michler), p. 21 (Dr. Morley Read); Zefa pp. 8, 11, 22.

Printed in the USA

1 2 3 4 5 6 7 8 9 99 98 97 96 95 94

At this time, Gareth Stevens, Inc., does not use 100 percent recycled
paper, although the paper used in our books does contain about 30
percent recycled fiber. This decision was made after a careful study of
current recycling procedures revealed their dubious environmental
benefits. We will continue to explore recycling options.

TABLE OF CONTENTS

Words that appear in the glossary are printed in **boldface** type the first time they occur in the text.

WHY IS WOOD USEFUL?

People have been using wood for thousands of years. Wood is strong, but it is easy to cut and shape. Wood is used to build ships, bridges, and houses. It has been used for centuries as a source of heat to keep us warm.

We have many objects made of wood in our homes, such as tables, chairs, and beds. Toys, and toy boxes to keep them in, are often made of wood. Wooden spoons are good for cooking because they don't get hot like metal spoons.

Many houses are built with wooden doors and floors. They may also have wooden **rafters** to hold up the roof.

➤ The inside of this house in Thailand is made of wood, and so is much of the furniture.

GROWING TREES FOR WOOD

Today, trees are often grown just for their wood. The wood they provide is either **hardwood** or **softwood**. Hardwoods, like oak, come from trees that grow slowly. Hardwoods are stronger and last longer than softwoods. They are used to make floors, furniture, and many other things.

◆ Violins are handmade from many kinds of wood.

Most trees grown specifically for their wood are softwoods. Softwoods, like pine trees and other **conifers**, grow more quickly than hardwoods.

Softwoods can be cut easily into different shapes by machines. Most softwoods are used to build objects and structures and to make paper for books, newspapers, and all types of packaging.

WOOD TRASH

Because wooden objects last a fairly long time, we don't usually throw much wood away. Some wooden things are thrown away because they are broken or damaged. But broken wood can be mended quite easily. A little glue and a few clamps, or sometimes a new paint job, will usually make broken toys and furniture look like new.

Most wood trash comes from **construction sites**. When old houses are torn down, their wood is not usually saved. The **timber** that is left over when new houses are built is also usually just thrown away.

⬆ Most of the wood from this torn-down house will probably be thrown away.

DUMPING WOOD

Wood waste that isn't burned is usually buried in a **landfill**. Wood is **biodegradable**. This means it is slowly broken down by **bacteria** until it decays and becomes part of the soil.

⬆ Tree trunks that fall on the ground slowly decay and enrich the soil.

Trash that is buried in a landfill is **compacted**, or pressed tightly together, by big compacting machines. Compacted trash cannot decay properly because it needs air and bacteria to break apart. So bulky wood trash that takes up a lot of space in the ground can quickly overload a landfill.

BURNING WOOD

Some people get rid of wood trash by
burning it. They burn old furniture and
other wood items. Burning wood
makes dirty smoke that **pollutes** the air.

⬆ Look at how much smoke this bonfire makes!

People sometimes burn the twigs and leaves they rake up in their yards in the fall. But leaves should not be burned. Instead, they should be **composted**.

HEAT AND GAS IN THE AIR

Fires need **oxygen** to burn properly. Fires also give off a gas called **carbon dioxide**. This gas keeps heat in Earth's atmosphere from escaping into space.

⬆ This wood fire is creating heat and carbon dioxide.

The heat trapped by the carbon dioxide warms up the air around Earth. This is called **global warming**. Global warming can have a powerful effect on Earth's weather and climate.

WHY WE NEED TREES

People and animals breathe in oxygen and breathe out carbon dioxide. Plants, however, take in carbon dioxide and release oxygen. So, when trees and entire forests are cut down, less oxygen and more carbon dioxide go into the air.

Trees are also useful because they protect the soil. Tree roots hold the soil in place, and leaves keep the soil from drying out. If too many trees are cut down, the rich soil **erodes**, or is worn away. Soon, nothing can grow on this land, and it becomes a desert.

⬆ The tree roots hold the soil in place, but the rest of the soil is badly eroded.

CUTTING DOWN FORESTS

Forests are being destroyed around the world. In many places, trees are burned to make room for roads or fields.

But forest soil isn't really meant for crops. After a few years, the soil is too weak to grow anything. Then more forests are cleared so people can grow crops again.

Bananas, coffee, and chocolate all come from plants that first grew in **rain forests**. Millions of other plants grow there as well. Some are useful in making medicines. Birds, insects, and other animals also live in these lush tropical areas. If we destroy all the rain forests, we will probably lose these plants and animals forever.

▲ Part of the Amazon forest is being cleared for farmland.

CUTTING DOWN WOODLANDS

Many **deciduous** forests are being cut down, as well as the rain forests. Some of the trees are lost when there is a storm or flood. But many trees are cut down so the land can be used for farming or for building houses and roads. Sometimes, deciduous forests are cut down so the land can be used to grow conifers for softwood.

◆ Deciduous trees change color in fall and then lose their leaves. Conifers keep their needles all year round.

Cutting **woodlands** also destroys the natural homes of delicate plants and forest creatures. Many animals make their homes under tree roots or build nests in the tree branches. Trees also provide food for some insects and other animals.

CARING FOR WOODLANDS

Woodlands should be cared for, not destroyed. Landowners can do this by removing diseased trees and thinning the trees so they aren't too crowded. This lets in more light, which helps **seedlings** and other plants grow better.

Many landowners used to **coppice** trees, and some foresters have begun to coppice trees again. This means they cut the tree trunk off to just above the ground. New **shoots** then grow out around it which can later be used to make fences and poles. This way, the tree doesn't die once its trunk is cut.

↥ The branches from these coppiced trees will be used to make poles and fence wood.

25

SAVING WOOD AND TREES

New trees are being planted in many countries. In Africa, for example, millions of seedlings have been planted to stop soil erosion and to provide wood for building and fuel. Also, some farmers are **interplanting** their crops. They grow crops around trees rather than cutting the trees down for planting.

◆ Here is a healthy corn crop interplanted among coconut trees.

Here is one way to help save trees: Collect tree seeds like acorns and plant them in moist soil in a pot indoors. Water them regularly, and soon you will have tree seedlings. When they are big enough, plant them outside in a spot that can use a new tree! Water them in dry weather and keep the soil around them free of weeds. You'll enjoy watching the trees grow from year to year.

GLOSSARY

bacteria: tiny creatures in the air, soil, and on plants and animals. Some bacteria break down dead matter or help the body digest food. Other bacteria can cause illness.

biodegradable: a substance that can be broken down by bacteria in the air.

carbon dioxide: a gas breathed out by people and animals and taken in by plants. Too much carbon dioxide in the air may eventually cause global warming.

compact: (v) to flatten and press together huge amounts of trash in order to make it take up less room in a landfill.

compost: (v) to heap together pieces of fruit, vegetables, and plants that have begun to decay. This matter slowly turns into a rich, soil-like substance. Compost helps make soil healthier for growing plants.

conifers: trees with needles or needlelike leaves and cones. Conifers do not lose their leaves in winter. Spruce, pine, hemlock, and fir are examples of coniferous trees.

construction sites: places where houses, businesses, schools, or other structures are being built.

coppice: to trim or cut a tree close to the ground so new shoots can still grow from the stump.

deciduous (trees): trees with leaves that change color and fall off in winter. Oaks and maples are examples of deciduous trees.

erode: to wear away a little at a time. Soil erosion often begins when trees are cut down.

global warming: the gradual warming up of Earth's atmosphere and climate because of too much carbon dioxide in the air.

hardwood: a strong, heavy wood taken from trees with broad, flat leaves. Hardwood trees, such as the oak, grow more slowly than softwood trees.

interplanting: planting a crop alongside or around another crop or plant. Farmers in the Philippines have tried interplanting corn and coconut trees.

landfill: a big hole in the ground where trash is dumped and then covered with soil.

oxygen: one of the main gases that make up air. People and animals need oxygen to breathe in order to stay alive.

pollute: to put harmful waste materials, such as gas, smoke, and trash, into the environment.

rafters: wooden planks used to support a roof or ceiling.

rain forests: forests in the hot, tropical parts of the world.

seedlings: newly sprouted plants; young plants grown from seeds.

shoots: sprouts; plant growths that have germinated from seeds.

softwood: wood taken from trees that have needles and cones, such as pine trees.

timber: wood used to build things.

woodlands: land that is covered with trees and shrubs.

PLACES TO WRITE

Here are some places you can write for more information about wooded resources and recycling. Be sure to give your name and address, and be clear about what you would like to know. Include a stamped, self-addressed envelope for a reply.

Greenpeace Foundation
185 Spadina Avenue
Sixth Floor
Toronto, Ontario
M5T 2C6

The National Recycling
   Coalition
1101 30th Street NW
Suite 305
Washington, D.C. 20007

Institute of Scrap
   Recycling Industries
1325 G Street NW
Suite 1000
Washington, D.C. 20005

INTERESTING FACTS ABOUT WOOD

Did you know . . .

▶ that the ships in which the Pilgrims first came to America were made out of wood? The sides of the ships were thickly coated with tar or pitch to keep out water.

▶ that types of trees and plants can become extinct, just like animals?

▶ that many musical instruments, from clarinets to violins to grand pianos, are made of wood because wood creates such a beautiful sound?

▶ that a wooden screw used to hold wood furniture together will work better than a metal one, because the wooden screw expands and therefore will rarely loosen up?

MORE BOOKS TO READ

Earthwise at Home: A Guide to the Care and Feeding of Your Planet.
  Linda Lowery (Carolrhoda)

In the Forest. Jim Arnosky (Lothrop)

Protecting Trees and Forests. F. Brooks (EDC)

Reducing, Reusing, and Recycling. Bobby Kalman (Crabtree)

Wood. J. Terry Jennings (Garrett Ed. Corp.)

VIDEO: *The Living Planet.* John D. and Catherine T. MacArthur Foundation
  Library Video Classics Project

INDEX